A Question of Math Book

Multiplication

by Sheila Cato
illustrations by Sami Sweeten

Carolrhoda Books, Inc./Minneapolis

Carolrhoda Books, Inc., c/o The Lerner Publishing Group
241 First Avenue North, Minneapolis, MN 55401 U.S.A.

Website address: www. lernerbooks.com

LIBRARY OF CONGRESS CATALOGING-IN-PUBLICATION DATA
Cato, Sheila
     Multiplication / by Sheila Cato : illustrations by Sami Sweeten.
          p.   cm. — (A question of math book)
     Summary: Introduces the concept of multiplication by presenting
simple problems taken from everyday life.
     ISBN 1-57505-321-7 (alk. paper)
     1. Multiplication—Juvenile literature. [1. Multiplication.]
     I. Sweeten, Sami, ill. II. Title. III. Series: Cato, Sheila, 1936-
Question of math book.
     QA115.C27 1999
     513.2'13—dc21   98-6377

The series A Question of Math is produced by Carolrhoda Books, Inc.,
in cooperation with Brown Packaging Partworks Limited, London, England.
The series is based on a concept by Sidney Rosen, Ph.D.
Series consultant: Kimi Hosoume, University of California at Berkeley
Editor: Anne O'Daly
Designers: Janelle Barker and Duncan Brown

Printed in Singapore
Bound in the United States of America

1 2 3 4 5 6 - JR - 04 03 02 01 00 99

This is Luis. He is learning about multiplication with the help of his number friend, Digit. Take a look at Digit. Can you see the numbers we use in math?

Luis's friends Brad, Holly, Josh, and Mia will be joining in the fun. You can join in, too. You will need counters, buttons, pennies, and colored pencils.

*My mom has just bought 2 packages of donuts.
There are 2 donuts in each package. I know
that I can add 2 and 2 together to give 4.
Is there another way of finding how many
donuts there are all together?*

Yes, Luis, there is. When you have two or more
groups with the same number of things in each
group, you can join them together in a special way.
It's called multiplication.

4

Special Signs

There are signs in the supermarket to show us where things are. Multiplication has special signs, too. The sign x means multiplied by or times. The sign = means equals.

You have 2 packages each with 2 donuts. We can write this as an equation.

$$2 \times 2 = 4$$

The first 2 is the number of packages. The second 2 is the number of donuts in each package. When you multiply two things together, the answer is called the product.

So 4 is the product of this equation and there are 4 donuts. Thanks, Digit.

Now You Try

If there were 3 donuts in each of the 2 packages, how many donuts would Luis have all together?

My cat has had some kittens. There are 3 kittens on the chair and 3 kittens on the floor. Can I use multiplication to find the total number of kittens?

Yes you can, Luis, because there are the same number of kittens in each group. There are 2 groups of kittens, and each group has 3 kittens in it. You can multiply the number of groups by the number of kittens in each group. The answer to that equation is the total number of kittens.

Number Sentence

When we write a letter to a friend, we put words into sentences. We can write about numbers in sentences, too. "Two multiplied by two equals four" is a number sentence.

2 multiplied by 3 makes 6.
So there are 6 kittens all together.

$$2 \times 3 = 6$$

I'd better give them some milk.

Now You Try

Write a number sentence to show how many collars Luis would need to buy if he bought a collar for each kitten.

Josh and I are making clown costumes to wear to a costume party. We are going to sew 4 buttons on the front of each costume. How many buttons will we need?

You will need 4 buttons for Josh's costume and 4 buttons for your costume. That's 2 groups of 4 buttons. Can you figure out the rest?

Six multiplied by one equals six

Sets and Groups

When we multiply two numbers, the first number tells us how many groups there are. The second number is how many things there are in each group. A group is sometimes called a set.

I could use my fingers. If I hold up 4 fingers on each hand and count how many fingers I am holding up, I get 8.

There's another way to figure it out, by solving the multiplication equation

$$2 \times 4 = 8$$

You get the same answer. Have fun at the party!

Now You Try

If Luis and Josh each put 3 buttons on their hats, how many buttons would they need?

9

Mia is helping me to hang my socks on the line. I have 5 pairs of socks.
If I use 1 clothespin for each sock, how many clothespins will I need?

This time, Luis, we have 5 sets because there are 5 pairs of socks. Does each set have the same number of things in it?

Each set has 2 socks. I could add all the socks together. 2 + 2 + 2 + 2 + 2 = 10.

But now that I know about multiplication, I can find the answer another way

$$5 \times 2 = 10$$

We will need 10 clothespins.

Keep on Adding

Luis found the answer to his problem by adding the socks 2 at a time. Multiplication is a quick way of adding the same number, over and over again.

Now You Try

How many clothespins would Luis need to hang 4 pairs of socks on the line? Remember – a pair is 2 of something.

11

8 clothespins

Our class is putting on a play. There are 5 fairies and 5 goblins in the play. How many children will get a part in the play?

The play sounds like fun, Luis, with lots of parts for you and your friends. First, let's think about how many sets there are.

12

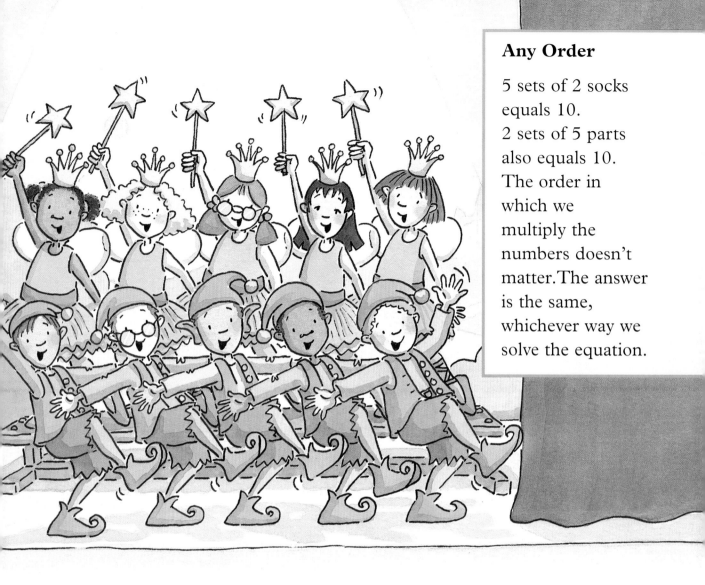

Any Order

5 sets of 2 socks equals 10. 2 sets of 5 parts also equals 10. The order in which we multiply the numbers doesn't matter. The answer is the same, whichever way we solve the equation.

There are 2 sets, a set of fairies and a set of goblins. Each set has 5 parts. We can find the total number of parts by solving a multiplication equation

$$2 \times 5 = 10$$

That means 10 children will play fairies and goblins.

Now You Try

If there were 6 fairies and 6 goblins, how many parts would there be in the play?

We're all going to summer camp for 2 weeks. How many days will we be away?

Well Luis, you know there are 7 days in a week. You want to find how many days there are in 2 weeks. Can you think of a way to find the answer?

12 parts

14

Multiplying by 1

If Luis was at summer camp for 1 week, he would be away for 7 days. 1 x 7 = 7. Multiplying a number by 1 means that the product is the same as the number.

That's easy! I can figure this out using multiplication. All I have to do is find 2 sets of 7.

$$2 \times 7 = 14$$

We will be at summer camp for 14 days.

Now You Try

Make a list of the days of the week. Cut out pictures from a magazine of your favorite foods and paste 3 pictures next to each day. How many pictures will you need?

Holly and I are making a paper chain using loops of colored paper. There are 2 colors, red and blue, and 8 pieces of each color. How many loops will that make, Digit?

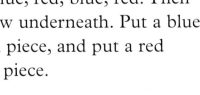

This is a good time to practice adding in 2s. Place 8 pieces in a row like this: blue, red, blue, red, blue, red, blue, red. Then make another row underneath. Put a blue piece under a red piece, and put a red piece under a blue piece.

Division

Multiplication joins equal sets to make one big set. Division separates one big set into smaller equal sets. Division is the opposite of multiplication.

Now count in 2s along the row – 2, 4, 6, 8, 10, 12, 14, 16.

There are 16 pieces of paper all together. I could have figured this out using multiplication

$$2 \times 8 = 16$$

Our paper chain looks great! Thanks for your help, Digit.

Now You Try

If Luis and Holly had 2 different colors of paper and 10 loops of each color, write an equation to show how many pieces they would have all together.

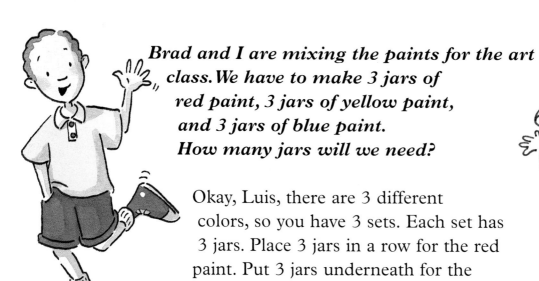

Brad and I are mixing the paints for the art class. We have to make 3 jars of red paint, 3 jars of yellow paint, and 3 jars of blue paint. How many jars will we need?

Okay, Luis, there are 3 different colors, so you have 3 sets. Each set has 3 jars. Place 3 jars in a row for the red paint. Put 3 jars underneath for the blue paint. What will you do next?

2 x 10 = 20

Odd numbers end in 1, 3, 5, 7, or 9. Even numbers end in 2, 4, 6, 8, or 0. When we multiply two odd numbers, the answer is odd. Two even numbers give an even number. When we multiply an even number by an odd number, the answer is even.

I'll make another row of 3 jars for the yellow paint and then I'll count all the jars.
There are 9 jars all together.
But it would be quicker to learn that

$$3 \times 3 = 9$$

We will need 9 jars in all.

Now You Try

Luis and Brad laid out 3 rows of jars, with 3 jars in each row. What do you notice about the shape the jars made?

We've arrived at the zoo in time to see the zookeeper feed the sea lions. There are 4 sea lions, and each one gets 4 fish. How many fish does the zookeeper have to bring in his bucket?

This is like your last question. The number of sets is the same as the number in each set. You can figure this out using buttons. Make 4 rows of buttons, with 4 buttons in each row, then count how many buttons there are all together.

The pots made a square. A square is a shape in which all the sides are the same length.

20

But the quickest way to figure this out is to learn your 4 times table. This tells you the answers you get when you multiply different numbers by 4.

$$1 \times 4 = 4$$
$$2 \times 4 = 8$$
$$3 \times 4 = 12$$
$$4 \times 4 = 16$$

The zookeeper needs 16 fish for the sea lions.

Square Numbers

When a number is multiplied by itself, the answer is a square number.

$1 \times 1 = 1$
$2 \times 2 = 4$
$3 \times 3 = 9$

1, 4, and 9 are all square numbers.

Now You Try

Use 9 red buttons to make a square. Add blue buttons so that there are 4 rows of buttons with 4 buttons in each row. How many blue buttons did you use?

We're helping to plant trees around 3 sides of the playground so there'll be some shade in the summer. We're going to plant 5 trees along each side. How many holes will we have to dig?

Let's start by thinking about 2 sides of the playground. There'll be 2 sets of 5 trees. You already know the answer to that, Luis.

To make sure that a multiplication answer is correct, add all the sets together. This should give you the same answer.

That makes 10. If I add the other set of 5 trees to 10, that makes 15. So we need to dig 15 holes for the trees.

$$3 \times 5 = 15$$

Digging is hard work – much harder than math!

Now You Try

The children planted 3 sets of 5 trees. Is 15 a square number?

Holly, Brad, Mia, and I are working on a dinosaur project together. We can each borrow 5 books from the library. How many books can we take all together?

There are lots of ways to solve this problem, Luis. There are 4 children – that's 4 sets each with 5 books. How would you write this as an equation?

15 is not a square number

24

Multiplying by 5

There's an easy way to check your answer when you multiply by 5. Even number x 5 – the last digit of the answer is 0. Odd number x 5 – the last digit of the answer is 5.

I write this as 4 x 5. I know that 2 x 5 = 10. Holly and Brad would have 10 books between them. Mia and I would have 10 books between us. I can find how many books we would all have together by adding 10 and 10. That makes 20, and I've found another way to figure out the answer.

$$4 \times 5 = 20$$

We'd better get to work!

Now You Try

If Josh had also been working on the project, how many library books could the children have borrowed?

My dad is building a patio in the back garden. There will be 5 rows of stones, and he'll put 5 stones in each row. How many stones will he need to buy?

Look at the plan, Luis. There are 5 rows of 5 stones. What have you learned about multiplying by 5? Can you see a pattern?

They could borrow 25 books

I know the answer must end in 5 or 0.
I have also learned

$$1 \times 5 = 5$$
$$2 \times 5 = 10$$
$$3 \times 5 = 15$$
$$4 \times 5 = 20$$

I think that 5 x 5 ends in 5 and begins with 2.
I could check the answer by counting
the stones in the patio.

$$5 \times 5 = 25$$

I was right!

Patterns

Look for patterns when you multiply. Will the answer be odd or even or a square number? Will it begin or end with a certain digit? Can you use what you already know to solve a problem?

Now You Try

Using graph paper, cut out a square of 6 rows with 6 squares in each row. Color the squares to make a pattern. How many squares are there?

We're decorating the school auditorium for a party. We're going to put 2 sets of colored lights across the stage. Each set needs 10 new bulbs. How many bulbs do we need?

There are 2 sets of colored bulbs, each with 10 bulbs. You want to find the answer to 2 x 10. Can you think of a way to figure this out?

28

Now You Try

Make a list of your age, the day, month, and year you were born, and your house or apartment number.
Multiply all these numbers by 10.

Since there are only 2 sets of 10, I could add 10 and 10 together. I know that I have 10 fingers, so I could just count all my fingers twice. That comes to 20.

$$2 \times 10 = 20$$

We need 20 new bulbs.

I know I have to learn multiplication tables. Is there a fun way to do this, Digit?

Look for patterns in this multiplication table. It might help you learn the multiplication facts.

X	1	2	3	4	5	10
1	1	2	3	4	5	10
2	2	4	6	8	10	20
3	3	6	9	12	15	30
4	4	8	12	16	20	40
5	5	10	15	20	25	50
10	10	20	30	40	50	100

All the numbers should end in 0

30

You can make your own multiplication table. Fill it in as you find the answers to multiplication questions. See if you can spot any patterns in the numbers.

The lines of numbers going across are rows.
The lines of numbers doing down are columns.
To find the answer to 4 x 2, look for the place where the row that starts with 4 meets the column that starts with 2. The answer is 8.

This multiplication table is a great way to find the answers to multiplication problems. I can look to see which numbers are in the table once and which are in twice. And I can look for all the square numbers.

31

Learning about multiplication is fun and it helps me figure out the answers to lots of problems. I can use it to find out how many donuts I've bought, how many paint jars I need, and how many light bulbs I should buy. I know that when we multiply a number by 1, the number stays the same. I can check my answer to see if I got it right. And I can look for patterns with square numbers and in a multiplication table.

Here are some useful multiplication words

Division:	This is a type of math that separates a big set into smaller sets of the same size.
Equation:	An equation is like a sentence in math. It uses numbers and special signs instead of words.
Even numbers:	Even numbers end in 2, 4, 6, 8, or 0.
Odd numbers:	Odd numbers end in 1, 3, 5, 7, or 9.
Multiplication table:	Multiplication tables tell you what you get when you multiply by 2, 3, 4, 5 and so on.
Product:	The product is the number you get when you multiply two numbers together.
Set:	A set is a group of things. Multiplication is about joining sets of the same size to make a bigger set.
Square number:	When you multiply a number by itself, the answer is called a square number.